ANIMALS IN THEIR WORLD

ECOSYSTEMS AND HOW THEY WORK

Julie K. Lundgren

A Crabtree Seedlings Book

CRABTREE
Publishing Company
www.crabtreebooks.com

Table of Contents

A Network of Life 4
Energy and Matter Move 8
Balance and Change 16
Glossary 23
Index 23

A Network of Life

Living and nonliving **matter** in an area form an **ecosystem**. Ecosystems contain water, air, minerals, plants, and animals.

desert ecosystem

water
animals
air
plants
minerals

The Sun provides energy to ecosystems. Different ecosystems have different conditions. To **survive**, each living thing has to have what it needs.

Sea lions hunt for fish in a kelp forest ecosystem.

Energy and Matter Move

Plants make food using water, minerals, air, and the Sun's energy through **photosynthesis**.

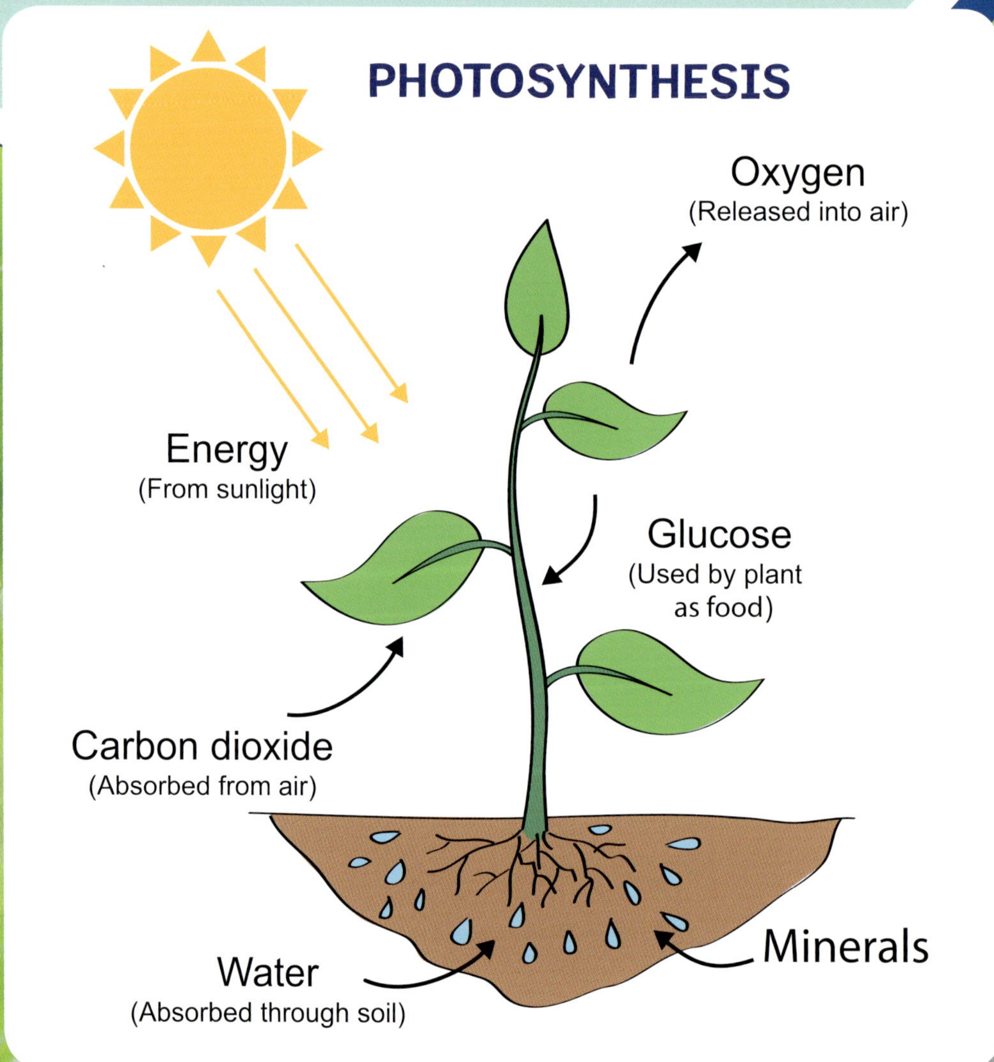

Plants are **producers**. They form the main part of ecosystems.

Animals are consumers—they eat to get energy. Herbivores eat plants. Herbivores may have body parts that help them eat certain plants.

Herbivores eat a lot to get the energy they need.

Special teeth chew and grind tough plants.

Carnivores eat other animals. Unlike plants, **prey** run or fight. Hunting takes a lot of energy.

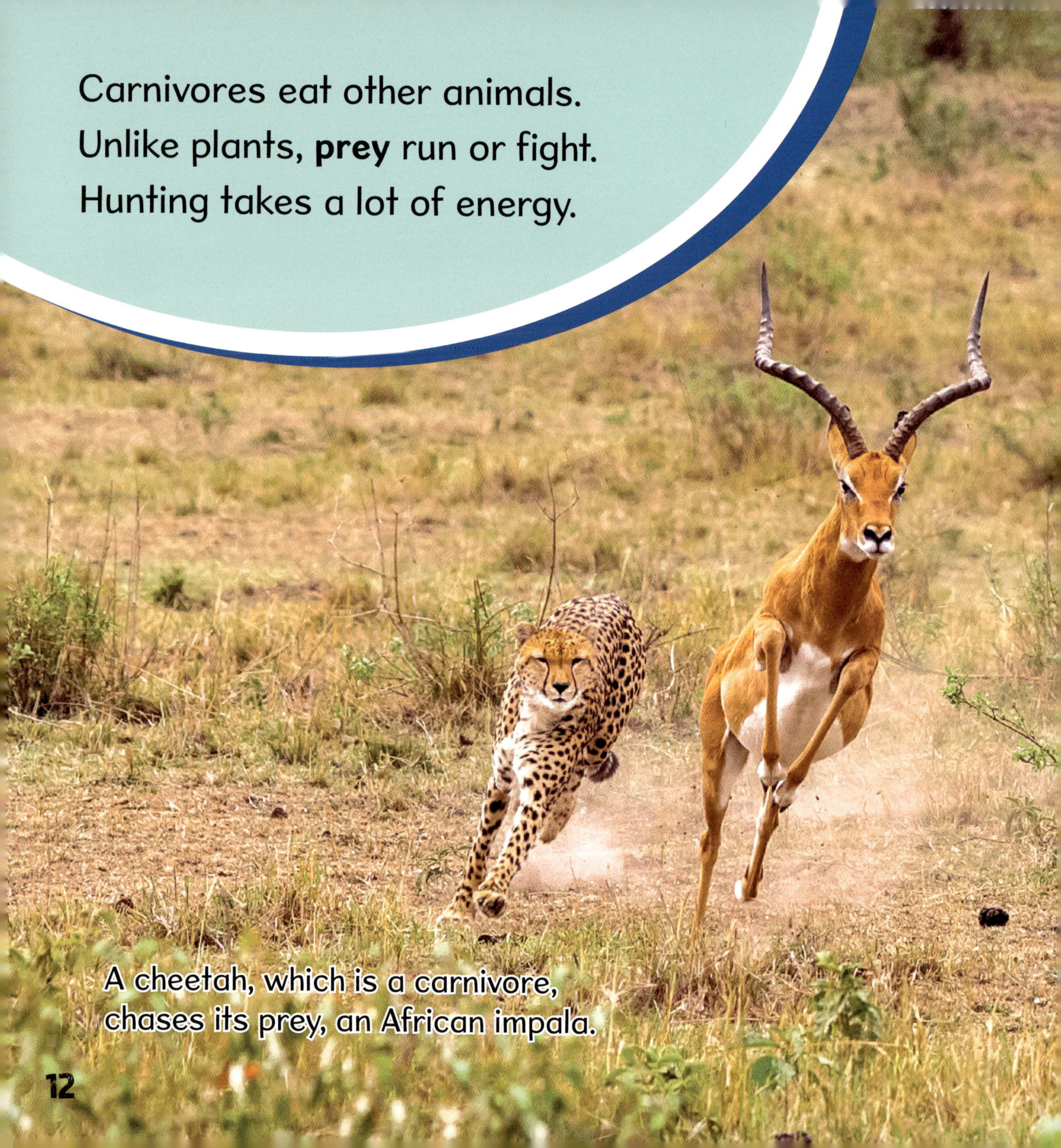

A cheetah, which is a carnivore, chases its prey, an African impala.

Animals use nonliving things in their ecosystem too, such as water and shelter.

A spotted owlet finds shelter in the forest ecosystem.

Decomposers recycle an ecosystem's leftovers. They break down animal and plant waste.

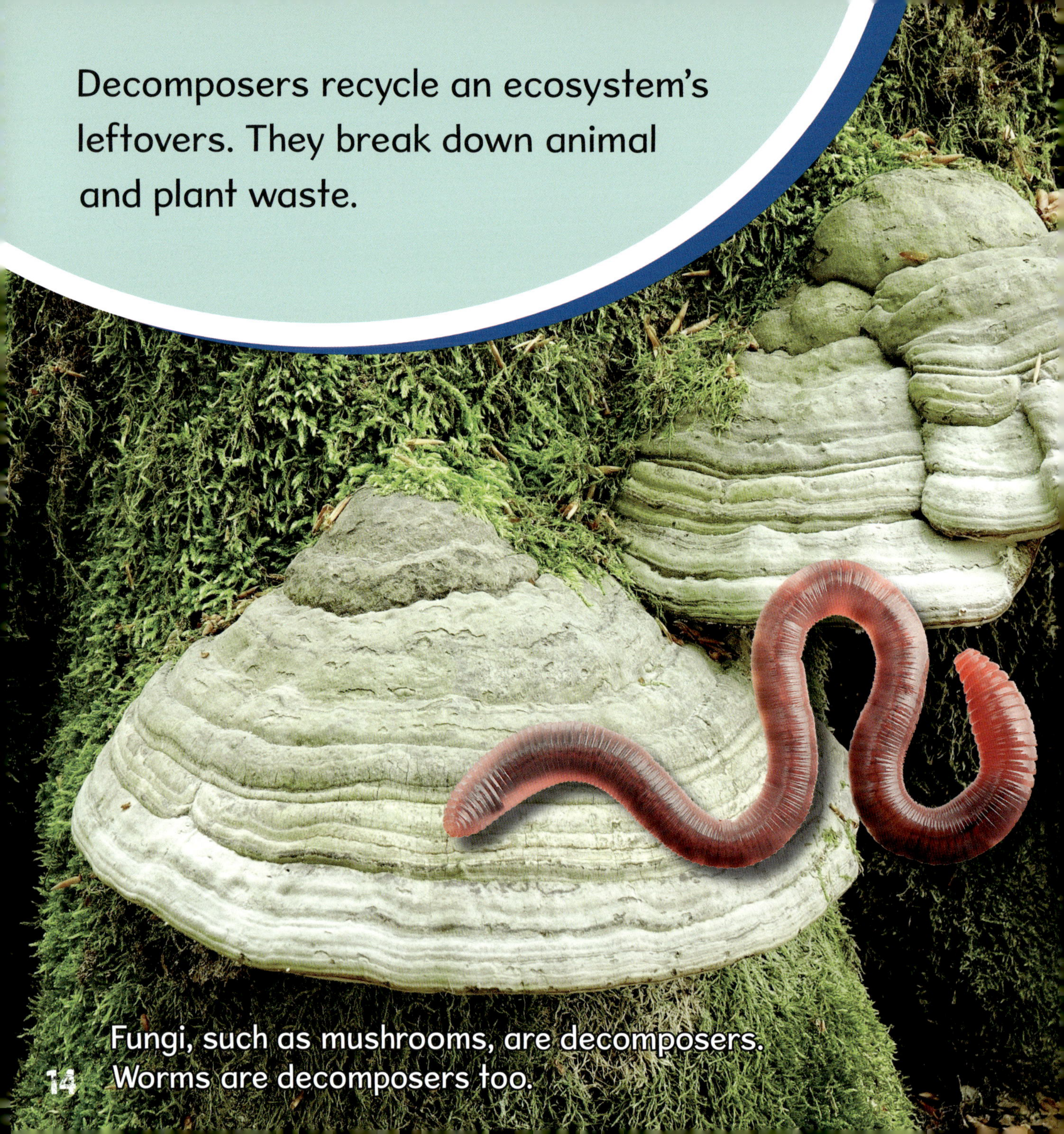

Fungi, such as mushrooms, are decomposers. Worms are decomposers too.

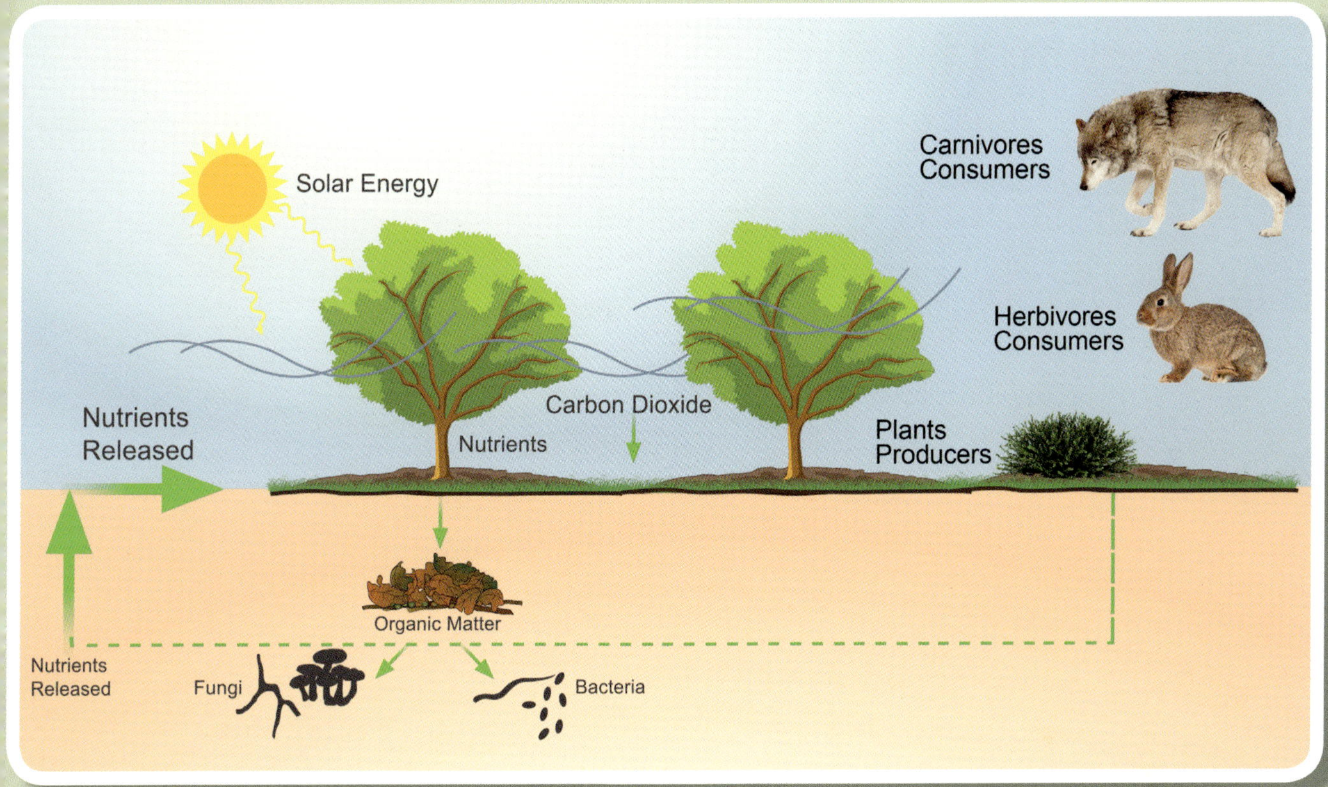

Energy and matter move through food webs.

Producers, herbivores, carnivores, and decomposers form a **food web**.

Balance and Change

Plants and animals in a healthy ecosystem support and balance each other year after year.

Small changes do not usually hurt a healthy ecosystem.

Big changes, or many changes, can upset the balance. Changes can be natural or caused by humans.

Severe weather can have a long-term effect on ecosystems.

Dry weather patterns can change ecosystems.

Human need for wood, fuel, food, and land must be in balance with the ecosystem for it to survive.

Help by cleaning trash and other unwanted items.

Protect healthy ecosystems. We are part of them too, and we need them to survive.

A red fox finds water outside of its natural ecosystem.

Climate change, growing cities, and increased mining and factory use affect Earth's ecosystems and us.

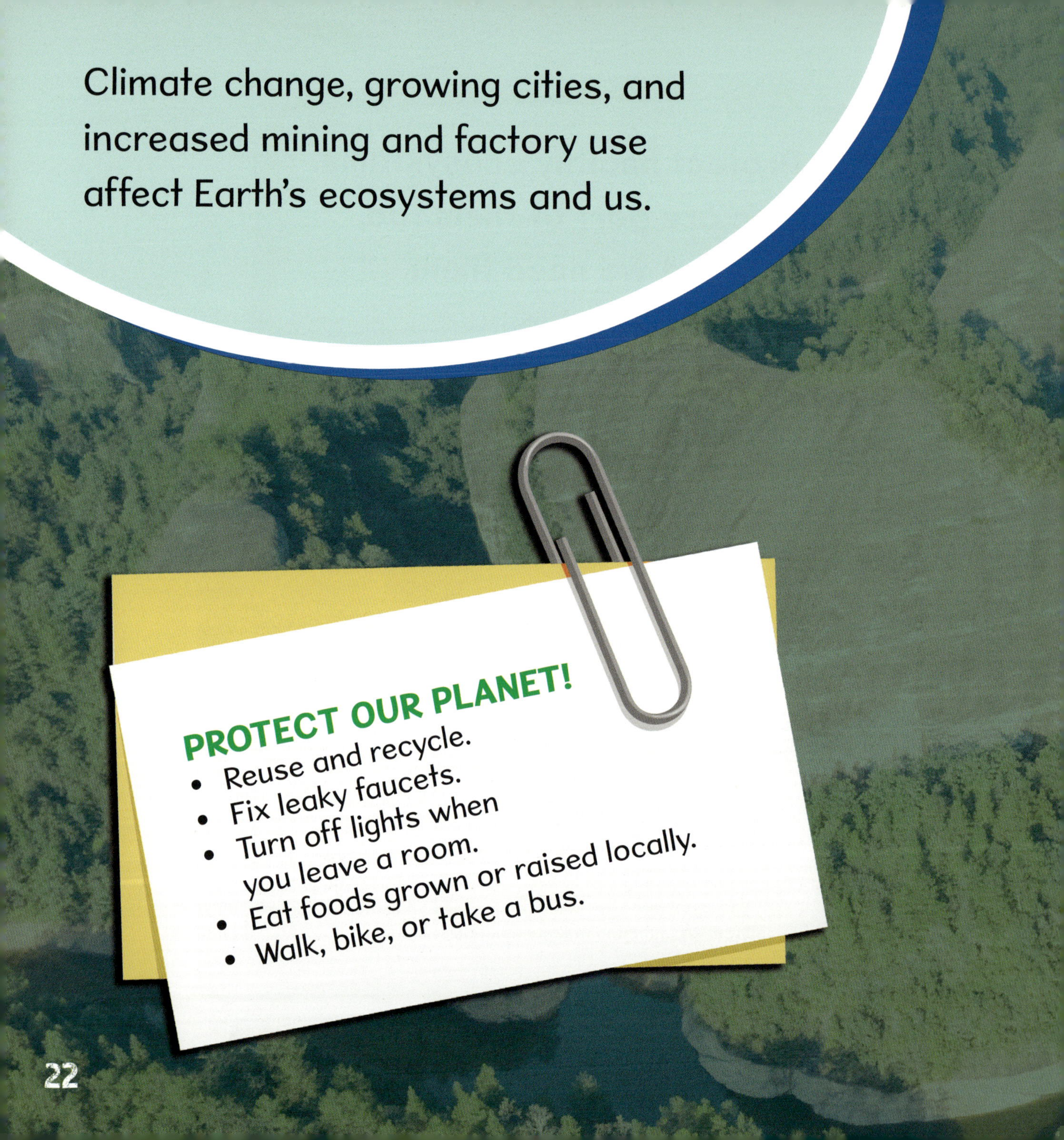

PROTECT OUR PLANET!

- Reuse and recycle.
- Fix leaky faucets.
- Turn off lights when you leave a room.
- Eat foods grown or raised locally.
- Walk, bike, or take a bus.

Glossary

ecosystem (EE-koh-siss-tuhm): A unit of nature that includes all the relationships between plants, animals, and the place they live

food web (FOOD WEHB): In an ecosystem, all of the links between living things based on what they eat

matter (MAT-ur): In an ecosystem, all living and nonliving things

photosynthesis (foh-toh-SIN-thuh-siss): The process by which green plants transform the Sun's energy into food

prey (PRAY): Animals that are eaten by other animals

producers (proh-DOO-serz): Plants that perform photosynthesis

survive (sur-VIVE): To continue to live, in spite of dangers

Index

balance 16, 18, 20
carnivores 12, 15
change(s) 17, 18, 19, 22
decomposers 14, 15
energy 6, 8, 10, 12, 15
herbivores 10, 15
matter 4, 15
plants 4, 8, 9, 10, 11, 12, 16
water 4, 5, 8, 13

School-to-Home Support for Caregivers and Teachers

This book helps children grow by letting them practice reading. Here are a few guiding questions to help the reader build his or her comprehension skills. Possible answers appear here in red.

Before Reading

- **What do I think this book is about?** I think this book is about how plants and animals together make an ecosystem. I think this book is about the many benefits of an ecosystem.
- **What do I want to learn about this topic?** I want to learn what animals live in different ecosystems. I want to learn what happens to the animals if an ecosystem is destroyed.

During Reading

- **I wonder why...** I wonder why plants are important in an ecosystem. I wonder why plant-eating animals need to eat large quantities of plants in order to survive.
- **What have I learned so far?** I have learned that worms are called decomposers because they break down animal and plant waste. I have learned that climate change and growing cities affect Earth's ecosystems.

After Reading

- **What details did I learn about this topic?** I have learned that I can help protect our planet by reusing and recycling materials and by eating food that is grown locally. I have learned that human need for wood, fuel, food, and land must be in balance with the ecosystem for it to survive.
- **Read the book again and look for the glossary words.** I see the word *survive* on page 6, and the words *food web* on page 15. The other glossary words are found on page 23.

Library and Archives Canada Cataloguing in Publication

Available at the Library and Archives Canada

Library of Congress Cataloging-in-Publication Data

Available at the Library of Congress

Crabtree Publishing Company
www.crabtreebooks.com 1–800–387–7650
Print book version produced jointly with Blue Door Education in 2023

Written by: Julie K. Lundgren
Print coordinator: Katherine Berti
Printed in the U.S.A./072022/CG20220201

Photo Credits:

www.shutterstock.com, www.istock.com ; Front Cover: Cover: shutterstock.com | Jag_cz. P2-3: shutterstock.com | Smileus. P4-5: istock.com | MriyaWildlife, istock.com | Björn Forenius, istock.com | amenic181, shutterstock.com | Harry B. Lamb, shutterstock.com | elenka_sergevna, shutterstock.com | Paul Aniszewski. P6-7: istock.com | VitalyEdush, istock.com | Michael Zeigler. P8-9: istock.com | PeJo29, istock.com | ipopba, istock.com | Ale-ks. P10-11: istock.com | Frizi, istock.com | PhotosbyAndy. P12-13: istock.com | vlapaev, istock.com | asxsoonxas. P14-15: istock.com | Omm-on-tour. P16-17: istpock.com | Joe Pearl Photography, istock.com | TahirAbbas, shuttterstock.com | Francisco Blanco. P18-19: shutterstock.com | Claudiovidri, shutterstock.com | TrustWolverines. P20-21: shutterstock.com | M_Agency P22-23: istock.com | chuyu

Published in the United States
Crabtree Publishing
347 Fifth Ave.
Suite 1402-145
New York, NY 10016

Published in Canada
Crabtree Publishing
616 Welland Ave.
St. Catharines, Ontario
L2M 5V6